Christelle Elvira Vioutou AIVODJI
Virgile AHYI

Study of total polyphenol levels

Christelle Elvira Vioutou AIVODJI
Virgile AHYI

Study of total polyphenol levels

And the variation in hydrogen potential (pH) of some honeys produced in northern BENIN.

Imprint
Any brand names and product names mentioned in this book are subject to trademark, brand or patent protection and are trademarks or registered trademarks of their respective holders. The use of brand names, product names, common names, trade names, product descriptions etc. even without a particular marking in this work is in no way to be construed to mean that such names may be regarded as unrestricted in respect of trademark and brand protection legislation and could thus be used by anyone.

Cover image: www.ingimage.com

This book is a translation from the original published under ISBN 978-620-6-69414-4.

Publisher:
Sciencia Scripts
is a trademark of
Dodo Books Indian Ocean Ltd. and OmniScriptum S.R.L publishing group

120 High Road, East Finchley, London, N2 9ED, United Kingdom
Str. Armeneasca 28/1, office 1, Chisinau MD-2012, Republic of Moldova, Europe
Printed at: see last page
ISBN: 978-620-7-20296-6

<u>**Dedication**</u>

- ❖ *My heart filled with immense joy, I dedicate this modest work first of all to my parents Joseph AIVODJI and Claudine BAH for their love, their sacrifices and their encouragement which made me what I am today. God grant you more of his love and preserve you in his blessings.*

- ❖ *My sincere gratitude to my sister Eunice AIVODJI for her wise advice and support; happiness and success in all areas.*
- ❖ *To my dear brothers Ulrich, Floris and Godson AIVODJI, thank you for everything you've done for me.*

<u>*Thanks*</u>

- ❖ *First of all, I would like to thank Almighty **God**, who gave me the health, strength, patience, courage and determination to accomplish this modest memoir. To you be glory; honor and adoration.*
- ❖ *To the members of the jury for their availability, the consideration and appreciation they accord our work in judging it, and the esteem in which we hold it.*
- ❖ *To Professor Virgile AHYI, you who represent our benchmark in scientific work. In the depths of our hearts, we give you credit for training, guiding and supporting us. May God protect you. Thank you for having spared neither your time nor your efforts to answer all our requests throughout our stay in your institute and in your analysis laboratory, precisely in the agro pharmaco analysis department.*
- ❖ *To IRGIB-Africa staff, teachers and administrators*
- ❖ *Laboratory staff, especially those in the agro pharmaco analysis department*
- ❖ *To all my colleagues in the class with whom I have shared the joys and difficulties of these years.*
- ❖ *To anyone who knows me and considers me a friend.*
- ❖ *To the people who encouraged and motivated me, who never stopped working for my success and happiness*

CONTENTS

List of abbreviations

Mg: milligram

GAE: Gallic Acid Equivalent

G: Gramme

pH: Hydrogen Potential

E: Sample

H+: Hydronium ion

Mr: Sir

Fe: Iron

 Cu: Copper

S: Suffering

NGO: Non-Governmental Organization

%: Percentage

Kg: Kilogram

ROS: Reactive Oxygen Species :

AND: Deoxribonucleic acid

RO-: Alkoxyl radical

 ROO-: Radical peroxide

B: Bèta

UV: Ultra Violet

IR: Infrared

Nm: Nano meter

λ : Lamda

A: Absorbance

T: Transmittance

e: extinction coefficient

l: length

dm: Decimeter

mol: Mole

Cm: Centimeter

c: concentration

H: Hydrogen

Cr: Chromium

O: Oxygen

H: Hydrogen

P: phosphorus

Mo: Molybdenum

W: Tungsten

N: Nitrogen

H.P.L.C: High Performance Liquid Chromatography

SM: Mass Spectrometry

N.M.R.: Nuclear Magnetic Resonance

ROS: Reactive oxygen species

+ : More

< : Lower and > : Upper

= : Equality

SUMMARY

Honey is considered a highly complex biological compound, endowed with a great diversity of properties, both nutritional and therapeutic. It is a foodstuff produced by honeybees from nectar or honeydew. In Benin, it is mainly produced in the north and sold in pharmacies, supermarkets, small stores in big cities, village markets and even at home. Much of the current research interest focuses on the study of antioxidant molecules of natural origin. The aim of our work is firstly to carry out a physicochemical study by determining the pH, followed by a phytochemical study by determining the polyphenols in our honey samples collected in northern Benin.

In this study, the quality of nine (09) samples taken in the septentrion was assessed. However, harvesting, packaging and storage conditions may influence quality. The Hydrogen Potential (pH) and polyphenolic compound content of honey samples were determined.

The physicochemical results obtained showed that the pH ranged from 3.2 to 5.12. Considering the pH values obtained in general, the limit permitted by the Codex Alimentarius 2001 standard is well respected, and there is a difference from one honey sample to another. As far as phytochemical results are concerned, the total polyphenol content assessed using the *Folin Ciocalteu* reagent as the assay method yielded a wide variability between samples (48.39 and 236.71 mg GAE / 100g honey). Sample E5 from Parakou is richer in polyphenols than the other samples studied.

Keywords honey, polyphenols, phytochemicals, physicochemicals

ABSTRACT

Honey is a very complex biological element with a great diversity which gives it many properties of both the nutritional and therapeutic standpoint. It is produced by honey bees from either nectar or honeydew. In Benin, honey is mainly produced in the north and marketed in pharmacies, supermarkets, small shops of major cities, market towns and even at home. Most of the current research activities focuses on the study of antioxidant molecules of natural origin. The aim of our work is to make first a physicochemical study based on pH determination and a phytochemical study based on the dosage of polyphenols in honey samples collected in the north of Benin.

In this study, the quality of nine (09) honey samples to take in the north of Benin were evaluated. However, conditions for harvesting, packaging and storage can influence its qualities. The Potential Hydrogen (pH) and proportion of polyphenolic in honey samples have been   determined.

The physicochemical results showed that the pH is ranged from 3.2 to 5.12, which is perfectly in accordance with the Codex Alimentarius in 2001 standards, with some variations from one honey sample to another. Regarding phytochemicals results, the proportion of polyphenols obtained while using *Folin Ciocalteu* as assay show a large variability between (48.39 and 236.71 mg GAE / 100 g of honey). The E5 sample from Parakou is richer in polyphenols compared with other samples.

Keywords Honey, Polyphenols, Phytochemicals, Physicochemical

INTRODUCTION

Honey is a naturally sweet substance produced by the beehive (Delphine Irlande, 2010). Given its nutritional qualities and therapeutic properties, it has been adopted by most ancient civilizations. Its organoleptic, physico-chemical and photochemical characteristics vary widely, depending on climatic and environmental conditions, as well as on the origin of the plants from which it is harvested (Cimpoiu et al., 2012).

Honey contains at least 200 substances, mainly carbohydrates and water. It also contains minerals, proteins, free amino acids, enzymes, vitamins, organic acids, flavonoids, phenolic acids and other phytochemicals (Terrab et al., 2003).

Antioxidants are endogenous or exogenous substances capable of neutralizing or reducing the damage caused by free radicals in the body. (DOUKANI Koula et al 2014). Studies relating to honey have revealed that its phenolic compounds contribute significantly to its antioxidant power.

Phenolic compounds are considered almost universal, forming the most important group of phytochemical compounds in plants (Beta et al., 2005). Structurally, they fall into several classes, ranging from compounds with a simple phenolic core (e.g. gallic acid) to complex polymeric compounds such as tannins (Ez-zohra, 2009). Moreover, several studies have shown that the antioxidant activity of honey varies widely depending on the floral source (Dimitrova et al, 2007); the greater the quantity of phenolic compounds, the better the antioxidant activity of honey (DOUKANI Koula et al 2014).

Oxidative stress is implicated in a wide spectrum of diseases that have an enormous impact on the health of populations. Under normal conditions, aerobic metabolism in mammals generates substances called reactive oxygen species (ROS), which are involved in small quantities in physiological processes (Favier, 2003). Ph, or hydrogen potential, is a measure of the potential characterizing the acidity or basicity of a medium. It represents the concentration of $H+$ ions in a solution. Their variation in honey is due to many factors.

Nowadays, with the rapid expansion of natural medicines and certain pathologies resistant to conventional treatments, honey can be an asset thanks to its therapeutic activities (Delphine Irlande, 2010). It is used in the treatment of cardiovascular disease, cancer, cataracts and many inflammatory diseases, as well as in wound healing. Honey's therapeutic actions are mainly due to its anti-oxidant and antimicrobial properties (Martos et al., 2000).

The aim of our work will be to evaluate the total polyphenol content of a few types of honey produced in northern BENIN and the variation in their ph, in order to assess the correlation between total polyphenol content and ph. Total polyphenols will be determined using the Folin-Ciocalteu spectrophotometer method, and the ph measured using a pH meter.

PART ONE

BIBLIOGRAPHIC SUMMARY

1. General

1.1. Bees

The bee is an insect belonging to the "Hymenoptera" family (RABEHARIFARA Zoelinoro Patricia 2011). It is one of the few insects to have been domesticated by man. Since ancient times, the products of insect breeding have played a key role in domestic and economic life, thanks to their multiple uses. As Voisenet points out in his work on the Christian bestiary, the bee is one of the few animals to possess an exclusively positive value (Voisenet 1994). What's more, the products derived from bee-keeping are unique in that they belong to both the animal and plant kingdoms (Catherine Mousinho and Marie-Christine Marinval).

They collect and feed on nectar and pollen, and play a tireless role in pollinating plants and maintaining the balance of nature. The honey bee belongs to the *Apis mellifera* species.

1.2. Taxonomy

The best-known and most studied bee is *Apis mellifera*. Its name means "which manufactures honey"(GUERZOU, 2002). Kingdom: Animal, it belongs to the Arthropods phylum, Class: Insects it belongs to the Aculeate Hymenoptera Order, Superfamily: Apoidea its Family: Apidae, Tribe: Apinae. With Genus: and Species: *Apis mellifera* it is of the Race: *unicolor.* The figure below shows this bee species:

Figure 1 *Apis mellifera* (RABEHARIFARA Zoelinoro Patricia 2011).

1.3. Some products from the hive

Honey: this is a natural sweet substance produced by *Apis* mellifera bees.
from the nectar of plants or from secretions from living parts of plants, or

from the excretions of foraging insects left on living plant parts, which the
bees gather, transform them by combining them with enzymes they secrete themselves, deposit, dehydrate, store and leave to refine and mature in the hive's combs (Codex *alimentarius*).

Figure 2: Honey (Oudjet kahina 2012)

Wax: The worker bee secretes wax as a building material for hive combs. Thanks to its four pairs of epidermal wax glands located on the inner surface of the last four abdominal rings, the 13 to 15-day-old bee produces wax in the form of small white scales (RABEHARIFARA Zoelinoro Patricia 2011). It is originally light yellow in color, but when used as a building material for hive cells, it becomes loaded with foreign matter that changes its color from yellow (due to the presence of a coloring agent: chrysin or 5-7 dihydroxyflavone) to dark brown (Alexandra ROSSANT 2011).

Figure 3 The wax **Figure 4** Small wax shavings

Alexandra ROSSANT 2011.

Pollen: Found in the hive, pollen contains B vitamins, proteins and trace elements (Fe, Cu, S) (RABEHARIFARA Zoelinoro Patricia 2011). However, for many years now, it has **been** used as a very interesting dietary supplement. Daily consumption of small quantities of pollen pellets can provide a very good supply of vitamins and make up for certain dietary deficiencies. Research on animals indicates that pollen can have a significant effect **on** growth (Alexandra ROSSANT 2011).

 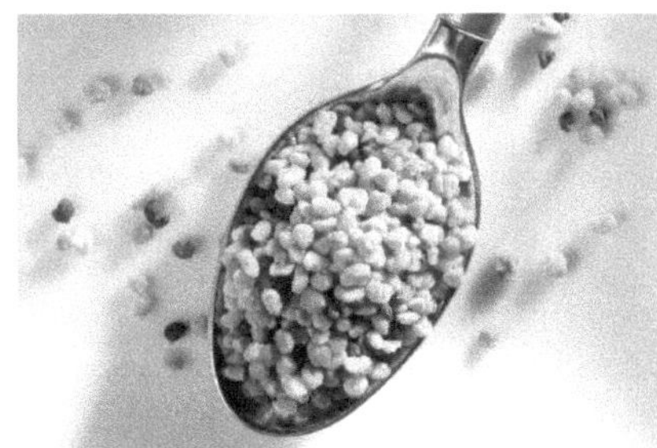

Figure 5 A pollen trap **Figure 6**: Pollen pellets
(Marchenay P. and Bérard L., 2007).

Royal jelly: a substance rich in growth factors, vitamins and minerals, very useful in pharmacy. Each hive contains a few milliliters) (RABEHARIFARA Zoelinoro Patricia 2011). Royal jelly is secreted by the pharyngeal and mandibular glands of young workers. It flows into the pharynx and accumulates in the crop before being deposited in the cells. It is the central substance of the hive, ensuring its existence and functioning. It is the queen's sole and exclusive food throughout her life. In fact, a queen can live for 5 to 6 years, compared with 45 days for a worker, and it's only the royal jelly diet that makes the difference. It's a whitish substance with a pearly sheen, gelatinous in consistency, warm-tasting, acidic (pH around 4) and very sweet (Alexandra ROSSANT 2011).

.

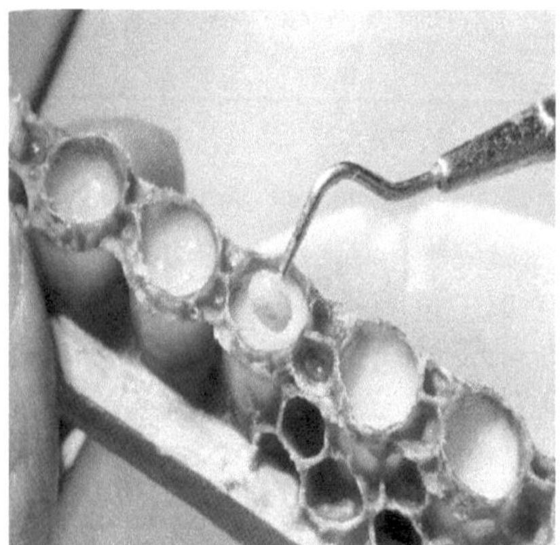

Figure 7 Harvesting royal jelly **Figure 8** Small jar of royal jelly (Alexandra ROSSANT 2011).

Propolis: This substance, harvested by bees from buds, is recognized as a treatment for cancer. Propolis **has** antibiotic properties (RABEHARIFARA Zoelinoro Patricia 2011). Propolis is used by bees for its multiple properties: it is a desiccant and antimicrobial substance used to clog the hive. Depending on its origin, the color of propolis varies from yellow to dark brown. The bactericidal properties of propolis have been known for centuries. Indeed, the combs of a beehive contain virtually no bacteria, as they are protected by a thin film of propolis (Alexandra ROSSANT 2011).

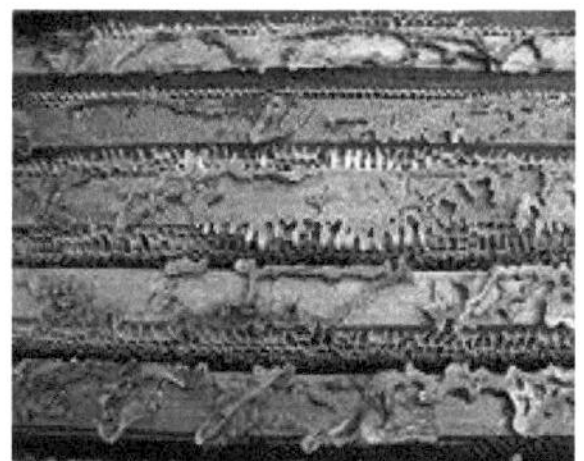

Figure 9 Bee carrying propolis **Figure 10** Propolis-covered frames Alexandra ROSSANT 2011

Bee venom: This is a plant product that can be extracted for its therapeutic action (RABEHARIFARA Zoelinoro Patricia 2011). Available from pharmacies in homeopathic form (*Apis mellifica*), bee venom is now used as a second-line treatment for disabling rheumatism, arthritis, tendonitis and certain autoimmune diseases, including multiple sclerosis (Lefief-Delcourt A., 2010). Venom is collected using a tray covered in nylon cloth, placed on a glass plate and positioned at the entrance to the hive. Any treatment using venom should always be carried out in hospital (Bruneau E., 2002). Bee venom is a mixture of proteins with a basic pH, whose bitter taste (isoamyl acetate) makes bees very aggressive. However, scientific data on this type of treatment are not yet sufficient (Soman Neelesh R. *et al.*, 2009).

Figure 11 A bee stinging nylon (Bruneau E., 2002).

2. <u>Honey</u>

2.1.<u>The different types of honey</u>

There are many varieties of honey, which can be classified as follows:

- Depending on the floral origin, there are two main varieties of honey, depending on secretory origin: nectar honey and honeydew honey.
- Depending on the geographical origin of the honey, which is based on pollen analysis we have single-flower honeys and multi-flower honeys. (Rabeharifara Zoelinoro Patricia 2011)

2.2.<u>Some honey produced in BENIN</u>

Honey is a foodstuff produced by honeybees from nectar or honeydew. In Benin, it is marketed and sold in pharmacies, supermarkets, small stores in big cities, village markets and even at home. However, the conditions under which it is harvested, packaged and stored can affect its qualities. Among these honeys, the most appreciated originate from the northern zone of Benin. Investigations carried out on samples of honey sold in Cotonou have revealed that some do not correspond to the geographical and botanical origins indicated by the packaging on the identification labels. Some producers or traders therefore abuse the image conveyed by honey designations to sell their products (J.A. Djossou, 2013).

In Africa, honey is widely consumed. Beekeeping is practised by private beekeepers or associations in different regions of the country. Most of the time, beekeeping is supported by non-governmental organizations (NGOs), development projects and charitable organizations, which in return buy the honey produced and package it for sale in towns and cities. Various names are often given, depending on where the honey is produced, or where the headquarters of these funding organizations are located, as they are more numerous in the north of the country.

Table 1 Designations of some honeys on the COTONOU market (Gbèkponhami Monique et al 2011)

N°	NAMES
1	Flora honey-Parakou

2	Honey from Donga
3	Tobe-Bantè honey
4	Honey from Alibori
5	Honey from Zou Nord
6	Gogounou GAMIA Bemberekè
7	INA Bemberekè
8	Beroubouay Bemberekè
9	Parakou
10	Kidaroukperou KALALE
11	Sekere SINENDE
12	Gnaro SINENDE
13	Kouande

2.3.Honey composition

Despite honey's complexity, its qualitative composition is fairly well known today. However, the proportions may vary. The table below details the general composition of honey:

Table 2 Composition of honey (Lavoisier, 1987)

Carbohydrates (sugars) 75-80%	Acids 0.3%	Proteins and amino acids 0.4%	Vitamins	Diastases	Minerals 0.2%	Miscellaneous
Reducing sugars Glucose: 31%. Levulose: 38 **Non-reducing sugars** Sucrose	**Gluconic acid** Succinic acid Malic acid Oxalic acid Glutamic acid Pyroglutamic acid Citric acid	**Materials albuminoids** Nitrogenous **matter** Traces of : Proline Trypsin Leucine Hystidine	Traces of : Thiamine: Vit B1 Riboflavin: Vit B2 Pyridoxine: Vit B6 Biotin: Vit B8 Ascorbic	Amylase α and β Invertase Traces of : Catalase Enzymes acidifying Glucose oxidase	Calcium Magnesium Potassium Iron Copper Manganese Boron Phosphorus Silicon	Esthers volatile Acetylcholine Pigments Colloids Factor antibiotic

5% maltose Isomaltose Erlose	Gluconic acid	Alanine Glycine Methionin e	acid: Vit C Pantothenic acid : Vit B5			(inhibin) Element s
	Formic acid Butyric acid Capric acid Caproic acid Valeric acid					
Meselzitose to Raffinose 10%. Kojibiose Dextrantriosi s	Butyric acid Capric acid Caproic acid Valeric acid	Aspartic acid	Folic acid: Vit B9 Nicotinamid e: Vit B3 = Vit PP			figures (pollen)

3. <u>Antioxidants : Polyphenols</u>

Many kinds of oxidants are produced and, despite their usefulness in the body, are also the cause of considerable damage.

To deal with these harmful oxidants, the human body possesses a whole arsenal of antioxidants.

But although the term "antioxidant" is often used, it is not easy to define, as it covers a large number of molecules and a wide variety of fields, including the food, chemical and pharmaceutical industries (Halliwell and Gutteridge, 1999). An antioxidant is any substance, present at a lower concentration than that of the oxidizable substrate, which is capable of retarding or preventing oxidation of this substrate. (Halliwell and Gutteridge, 1999). Antioxidants are powerful compounds capable of neutralizing the free radicals involved in cellular degradation, and enabling us to maintain an active, healthy lifestyle. Some antioxidants are manufactured by the human body, while others, such as vitamins and polyphenols, must be supplied by our diet. (Pincemail and Defraigne, 2004).

They represent a very varied group of substances. We can cite
flavonoids, tannins, phenylpropanoid derivatives such as lignans, esters
and hydroxybenzoic amides, stilbenes, coumarins and hydroxybenzoic acids,
xanthones and new compounds are continually being identified (Maarouf, 2000);
Although highly diversified, they all have one or more of the following in common

benzene rings bearing one or more hydroxyl functions Hopkins, 2003 ;.

Polyphenols are divided into several categories (see Georgé *et al.*, 2005).

- Simple phenols (C6): a single phenol nucleus, as for acids
phenolics (C6-C1).
- Flavonoids (C6-C3-C6): 2 aromatic rings linked by a heterocycle
oxygenated.
- Hydrolysable and non-hydrolysable tannins.
- Stilbenes (C6-C2-C6).
- Lignans, lignins and coumestans: 2 phenylpropane units

- Other phytoestrogens
- Saponins (triterpenoids)
- Phytosterols and phytostanols (Paraskevi, Moutsatsou, 2007). Although they do not
are not polyphenols, isothiocyanates are usually added to this list,
    derived from the hydrolysis of glucosinolates (Dacosta, 2003).

3.1 <u>Free radicals and oxidative stress</u>

Oxygen is an invisible, odorless gas that is indispensable to multicellular organisms, because it enables them to produce energy by oxidizing organic matter. But our cells transform some of this oxygen into toxic metabolites: organic free radicals (Lesgards, 2000). A free radical is a species, atom or molecule, containing an unpaired electron. This imbalance is only transient, and is filled either by the acceptance of another electron, or by the transfer of this free electron to another molecule. These highly unstable and reactive radical species are produced continuously in our bodies, **as** part of many biological phenomena. For example, during cellular respiration, molecular oxygen is transformed into various oxygenated substances, commonly known as oxygen free radicals or Reactive Oxygen Species (ROS) (Gutteridge, 1993). In certain situations, this production increases sharply, leading to stress. defined as an imbalance between the production and destruction of these substances. species (Gutteridge, 1993). This imbalance is due to a number of factors, including pollutants in the air we breathe and the water and food we consume. Ultraviolet rays from the sun, other radiation, tobacco smoke and excessive exercise are also factors that considerably increase the presence of free radicals in our system (Favier, 2003). Given their ability to damage cells, tissues and organs, reactive oxygen species are implicated in a wide range of pathologies, both acute and chronic (Gutteridge, 1993). Free radicals are mainly produced by endogenous sources, such as electron transport chains, peroxisomes and the cytochrome P-450 system. These radicals are responsible for DNA damage and cellular aging, which are at the root of diseases such as atherosclerosis, cancer, Alzheimer's and Parkinson's (Favier, 2003). It is not easy to classify all known antioxidants capable of fighting free radicals; they are generally classified according to their mechanism of action or their chemical nature.

3.2.<u>Mechanism of action</u>

<u>Classification of antioxidants by mechanism of action</u>

Regardless of where they are located, antioxidants can act on two levels: preventing the formation of oxygenated free radicals or purifying free radicals oxygenated. In addition to this double line of defense, the body is also capable of repairing or eliminating molecules damaged by radical attack.

<u>Group I</u>

Various names have been given to this group: free radical scavengers, primary antioxidants, chain breakers. These antioxidants inhibit the initiation and propagation of oxidation by participating in the oxidation process and converting free radicals to their inactive forms. Primary antioxidants are generally phenolic compounds (AH) capable of donating a hydrogen atom to the free radical and converting it into a stable, non-radical compound. Group I antioxidants react predominantly with peroxylated radicals, for two reasons: the high concentration of these radicals and the low energy of the (ROO-) group, compared with other radicals such as (RO-), and the low concentration of the free-radical scavenger in the food. A free radical scavenger, even at low concentrations, competes with lipids to render the free radical inactive via an electron release reaction, followed by deprotonation.

<u>**Group II**</u>

They are classified as preventive compounds or secondary antioxidants. The latter encompass a range of different chemical substances which inhibit lipid oxidation by different mechanisms and do not transfer the free radical to its non-radical form. Very often, secondary antioxidants are linked to the inhibition of oxidation-initiating factors. Group II includes: pro-oxidative metal chelators, singlet oxygen deactivators, oxygen molecule scavengers, pro-oxidative enzyme inhibitors, antioxidant enzymes and hydroperoxide destroyers. For example, ascorbic acid can be a free-radical scavenger, deactivator of singlet oxygens in aqueous solution and effectively regenerator of tocopherol. Several flavonoids are free radical scavengers and metal chelators. (Kanoun Khadidja 2011)

3.3. <u>**Some therapeutic benefits of honey's antioxidants**</u>

Honey's therapeutic actions are due to its antioxidant and antimicrobial properties. The polyphenols contained in honey are responsible for its beneficial effect on health. Indeed, its role as a natural antioxidant is attracting increasing interest in the prevention and treatment of cancer, inflammatory, cardiovascular and neurodegenerative diseases, and skin wounds.
In today's world of natural medicine, and in the face of certain pathologies that are resistant to the usual treatments, honey can be an asset thanks to its therapeutic activities. Some of these therapeutic effects have since been confirmed. It can be described as antianemic, antiseptic, aperitive, bechic, digestive, diuretic, dynamogenic, emollient, febrifuge, laxative, sedative and vicarious. In addition, its nutritional qualities make it beneficial for both healthy and sick people. In addition to these general properties, each honey combines the medicinal virtues of the dominant flower from which it comes.

In the body, free radicals **are** responsible for cellular damage linked to
aging, cardiovascular disease, cancer and most other illnesses.
degenerative processes. Within the body, they are generated by lipid metabolism. The lower the level of oxidants, the lower the risk of cancer.
Certain foods contain antioxidants that are beneficial for the protection of the environment.
or the prevention of certain cancers, as is the case with honey. Plants contain numerous polyphenol-type molecules (Delphine Irlande 2010).

Among the many diseases mentioned above, we will focus on a few:

- ❖ Honey has a preventive action against cancer: the risk of cancer is increased in a population that is deficient in vitamins, trace elements and certain nutrients essential for cell metabolism and the production of enzymes and hormones. Most of these nutrients, which have an inhibitory effect on the carcinogenic process, are found in honey. These include flavonoids, which have the capacity to slow down the evolution of tumors.

- ❖ Honey regulates heart function: honey's sugar content makes it an excellent fuel for the heart muscle. What's more, the acetylcholine in honey slows and regulates the heartbeat, helping to lower blood pressure and improve blood circulation in the coronary arteries.

- ❖ Honey helps balance the neurovegetative system: Honey contains a wide range of minerals, vitamins and many other trace elements, the quantities of which vary according to the presence of pollen grains. This combination, which creates the conditions for the proper regulation of metabolic processes, also helps maintain a balanced nervous system.

- ❖ Wound healing: Its antibacterial and healing action is certainly best used in the external treatment of wounds.

- ❖ has antioxidant properties: Honey's antioxidant protective mechanism involves enzymes such as catalase and peroxidase, phenolic compounds, flavonoids, organic acids such as ascorbic acid and amino acids such as proline (Meda A. *at al.*, 2005). However, phenolic compounds are the most important in this activity. Antioxidants are substances which, present in low concentrations, are capable of suppressing, delaying or preventing oxidation processes and their consequences. Sources of antioxidants are many and varied: herbal extracts, honey, fruit, vegetables and tea. Antioxidants are classified according to their mechanism of action:

- primary antioxidants or anti-free radicals (type I): their actions are based on their ability to inactivate free radicals, as they inhibit the propagation of free radical reactions by supplying hydrogens to the free radicals present. Phenolic compounds belong to this class.

- Secondary or preventive antioxidants (type II): prevent the formation of free radicals by various mechanisms. Some chelate metal ions, reducing the pro-oxidant effect of ions, as in the case of certain organic acids and proteins. Others are oxygen scavengers, such as ascorbic acid, β-carotenes or certain enzyme systems.

NB: Some phenolic compounds have both type I and type II modes of action (Genot C. *et al.*, 2004). As a general rule, dark honeys and honeys with a high water content have a greater antioxidant capacity than other honeys. What's more, the antioxidant activity of honeys varies widely from one to another, and depends essentially on their botanical origin.

PART TWO

MATERIALS AND METHODS

PART TWO: MATERIALS AND METHODS

1. Hardware s

1.1. pH meters

Figure 12 Electronic pH meters

Figure 13 Manual pH meters

1.2. Spectrophotometer

Figure 14 Spectrophotometer

1.3. Physico-chemical analysis: pH determination

1.3.1. pH-metry: pH measurement

This is the determination of the pH of a solution. The pH value is used to characterize the nature of a solution, whether acidic (<7), basic (>7) or neutral (=7). It is measured using a manual or electronic pH meter. The pH is a criterion of honey quality that is included in international standards. The pH of honey is acidic, fluctuating between 3 and 6 (Clémence HOYET 2005).

1.3.2. pH meter operating principle

When the probe of a pH meter is immersed in an aqueous solution, an electrical voltage U appears across the electrodes, due to an electrochemical cell phenomenon. When equilibrium is reached between the study solution and the probe (permanent magnetic agitation of the study solution is necessary to achieve equilibrium), this voltage U is a decreasing affine function of pH: $U = a - b.\,\mathrm{pH}$

Where a and b *are* positive coefficients which depend on the nature of the electrodes, the solutions in which they are immersed and the temperature. The values of constants a and b are adjusted by calibrating the pH meter.

1.3.3 Using a pH meter

A pH meter consists of two parts: a probe with two electrodes immersed in the aqueous solution whose pH is to be measured, and an electronic unit connected to the probe, which displays the pH value. There are three stages in the use of a pH meter: temperature calibration; pH calibration; pH measurement (Pierron 2008).

1.4. Phytochemical analysis: Determination of total polyphenols

1.4.1. Spectrophotometry

Of the many instrumental methods used to determine the concentration of a chemical species in solution, the most common are those based on measuring the intensity of absorption or emission (spectrophotometry) of electromagnetic radiation by the species to be measured.

The most frequently used types of radiation are ultraviolet (UV), visible light and infrared (IR) (1st Year of Pharmacy 2002-2003).

Spectrophotometry can be used to :

- Determine the concentration of a colored chemical species in solution from its absorbance.
- Follow the kinetics of a slow chemical transformation...

1.4.2. **Principle of spectrophotometry**

Spectrophotometry is a technique for the qualitative and quantitative analysis of substances absorbing electromagnetic radiation at wavelengths between 300 and 900 nm, depending on the type of instrument used. When a substance absorbs in the visible range (400 nm $< \lambda <$ 700 nm), the eye perceives only the light that is visible.
unabsorbed radiation, which is why it appears colored, in the color
complementary to that of the absorbed write-off (Philippe Morin).

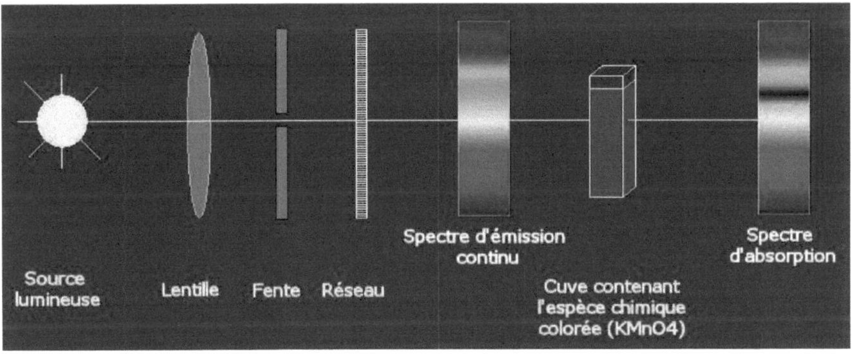

Figure 15 Operating principle of a spectrophotometer

1.4.3. **Beer-Lambert law**

Light absorption is directly proportional to both the concentration of the absorbing medium and the thickness of the cell containing the medium.
A combination of these two laws (the **Beer-Lambert** law) gives the relationship between absorbance (A) and transmittance (T):

A = log (I0 / I) = log (100 / T) = e c l

with :

A = absorbance (without unit)

e = molar absorption or extinction coefficient (dm3 . mol-1 . cm-1)

c = molar concentration (mol . dm-3)

l = tank length (cm) or light path

It is important to note that e is a function of the wavelength, and so the law of

Beer-Lambert is only true in monochromatic light.

The relationship between transmittance and concentration is not linear (*Figure 5),* but the

relationship between absorbance and concentration is linear (*Figure 6*), which is the basis for

the

most quantitative analyses (Process Engineering).

> **<u>Conditions for the validity of the Beer-Lambert law</u>**

- The light used is ***monochromatic***
- The concentration is not too high: $c \approx 10^{-2}$ mol.L-1 in general so that interactions
  between molecules are negligible.
- The solution does not fluoresce: no re-emission of light in all directions
- The solution is not too concentrated in colorless salts
- Dilution does not shift the chemical equilibrium:

$Cr2O_7^{2-}$ (orange) + H2O $- - \rightarrow$ 2 HCrO $_4^{-}$ (colorless in acid medium)

- The solution must be clear (no precipitate or cloudiness that would cause light
  scattering) (- C. Baillet TP Spectrophotométrie.doc)

2. <u>Methodology</u>

Ph will be determined using a pH meter. Then the quantitative analysis of total
polyphenols will be performed spectrophotometrically using Folin Ciocalteu reagent as
assay method (DOUKANI Koula1 et al 2014).

2.1. <u>Determining pH</u>

The pH of a solution is measured by a pH meter, which indicates the difference in electrical potential between a reference electrode with a fixed potential and an electrode whose potential varies with pH (glass electrode). These two electrodes must be immersed in the solution to be studied, and the electrical potential measured gives the pH value. It should be noted that the two electrodes are generally sold in the form of a single electrode (1ère Year of Pharmacy 2002-2003).

2.1.1. Protocol

The pH is measured in a 10% honey solution using a pH meter (Codex Alimentarius, 2001).

To measure the pH, you need :

- Place the solution to be analyzed under magnetic stirring.
- Dip the clean, dry probe into the solution to be analyzed, after calibration of course (position the probe carefully to avoid any impact with the probe);
- Wait for pH value to stabilize before reading.
- Read the pH value on the electronic unit connected to the probe

NB: Between two measurements, or between two calibrations, or between a calibration and a measurement, the probe must be washed in distilled water and dried with absorbent paper.

2.1.2. Equipment

pH meter, absorbent paper, beaker.

2.1.3. Reagents

Distilled water, calibration buffer, 10% honey solution

2.2. Determination of total polyphenols

The Folin Ciocalteu method was chosen for the determination of polyphenols for the following reasons: it meets the criteria of feasibility and reproducibility, offers availability of the Folin reagent and is well standardized. The chromophore's long absorption wavelength (760nm) minimizes interference with the sample matrix, which is often colored.

2.2.1 Principle

Total polyphenols in extracts of different types of honey were determined spectrophotometrically using the Folin Ciocalteu method (Singleton and Rossi, 1965), which consists of a mixture of phosphotungstic acid ($H_3PW_{12}O_{40}$) and phosphomolybdic acid ($H_3PMo_{12}O_{40}$), which is reduced during phenol oxidation to **a** mixture of blue oxides of tungsten and molybdenum. The blue color produced is proportional to the quantity of polyphenols present in the plant extracts, whose maximum absorption can be measured at 750 nm or 760nm.

Gallic acid is used as the standard and results are expressed as mg gallic acid equivalent per 100 grams of honey (mg GAE/100g), using the linear regression equation of the calibration curve plotted for gallic acid.

2.2.2. <u>Protocol</u>

To determine phenolic compound content. Five grams of honey was diluted with 50 ml distilled water and filtered. This solution (0.5 ml) was then mixed with 2.5 ml of 0.2 N Folin-Ciocalteu reagent for 5 min and 2 ml of 75 g/l sodium carbonate (Na_2CO_3) was then added. The absorbance of the reaction mixture was measured at 760 nm against a methanol blank after 2 hours. Gallic acid was used as the standard.

2.2.3. <u>Equipment</u>
UV visible spectrophotometer, volumetric flask, filter paper, beaker.

2.2.4. <u>Reagents</u>

Distilled water, Folin-Ciocalteu reagent, sodium carbonate, methanol, gallic acid.

PART THREE

RESULTS AND DISCUSSION

1. Results and discussion of physicochemical analyses

1.1.Hydrogen potential (pH)

Figure 13 shows the pH of the honeys obtained per sample.

The pH of the honeys studied ranged from 3.2 to 5.12, with an average of 4.44, so all the honeys analyzed were judged to have an acidic character.

Considering previous studies showing that nectar honeys have a pH between 3.5 and 4.5, and honeydew honeys between 5 and 5.5, we note that our results conform to this average. Considering the average, our samples studied do not exceed the permissible limit; we conclude that our samples are nectar honeys. However, our results clearly show that samples (E1, 2, 3, 5, 6 and 8) with pH values of 3.2 - 4.30 - 4.15- 4.21- 4.19- 4.62 respectively are between 3.5 and 4.5, and are therefore nectar honeys. On the other hand, the samples (E4, 7,9) with pH values of 5.10- 5.12 -5.10 are between 5 and 5.5 and are therefore honeydew honeys.

The variation in pH would be due to the flora foraged, the bee's salivary secretion and the enzymatic and fermentative processes during raw material processing. In view of the results obtained, 6 of the samples considered are of the nectar type and 3 of the honeydew type. The pH values therefore vary according to the sample. This enabled us to identify the two best-known types of honey.

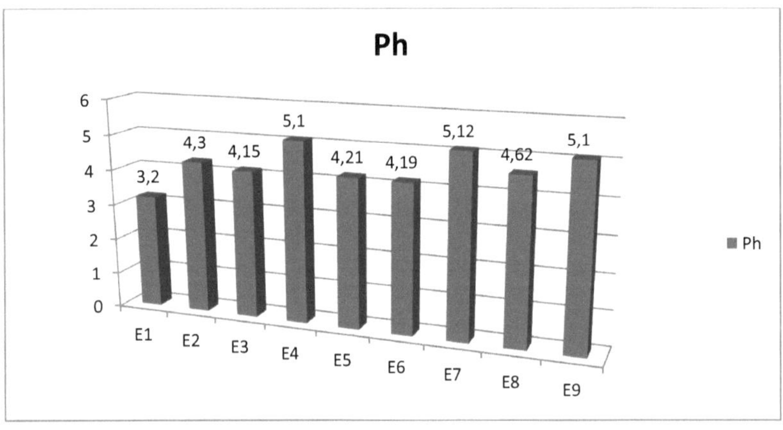

Legend

pH: y-axis

Honey sample (E): X-axis

2.   Results and discussion of phytochemical analyses

2.1. Total polyphenols

Polyphenols are one of the most important groups of natural compounds of high therapeutic interest. They were measured in honeys using gallic acid (R2=0.9943; y=0.029x-0.0107).

The determination of total polyphenols gives us an overall estimate of the content of different classes of phenolic compounds in the samples analyzed. The results obtained show that the concentration of polyphenols recorded in the honeys varies considerably from 48.39 to 236.71 mg GAE / 100g honey. These values vary according to honey type, as shown in figure 17.

The lowest value was recorded in honey (E4) (48.39 mg GAE /100g honey) and the highest concentration of polyphenols was established at (236.71 mg GAE /100g honey) for the sample (E5) suggesting that it has a better antioxidant potential.

In honey, most phenolic compounds are in the form of flavonoids, whose concentration depends on various factors, including the plant species used by bees, plant health, season and environmental factors. Variation in polyphenol content is due to the geographical origin of honey. Polyphenols are known for their antioxidant and biological virtues. They help prevent degenerative and cardiovascular diseases. Polyphenols are capable of scavenging the free radicals constantly generated by our bodies. They are involved in the prevention of cancerous diseases. Sample (5) is richer in polyphenols than the others, has a higher antioxidant content and therefore possesses enough antioxidants to help trap free radicals in the body.

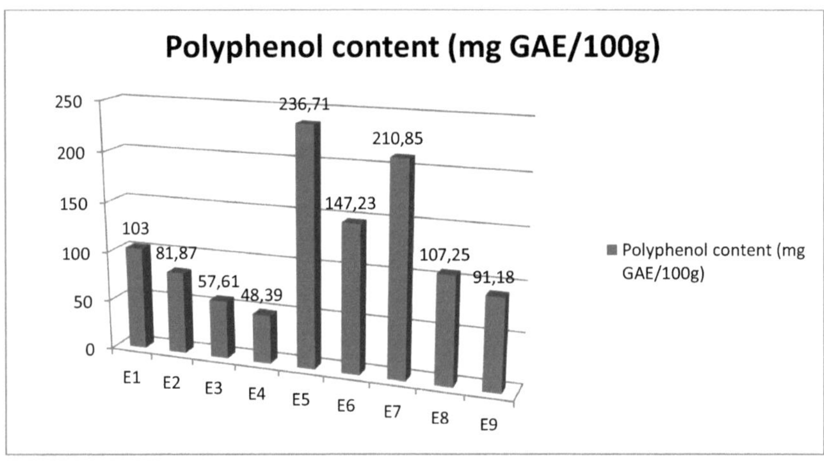

Figure 17: Total polyphenol content of honey samples.

Caption:

Polyphenol content: On the ordinate

Honey sample (E): X-axis

Table 3: Summary of results obtained

Sample No.	Name of honey (Municipality and village of origin)	pH	Total polyphenols (Mg/100g)
E1	Indicator	3,2	103
E2	Gogounou Gamia Bemberekè	4,30	81,87
E3	Ina Bemberekè	4,15	57,61
E4	Berebouy Bemberekè	5,10	48,39
E5	Parakou	4,21	236,71
E6	Kidaroukperou Kalale	4,19	147,23

E7	Wassa pehounko Sekere Sinende	5,12	210,85
E8	Gnaro Sinende	4,62	107,25
E9	Kouande	5,10	91,18

Conclusion

This study provided scientific data on the physicochemical (pH) and phytochemical (polyphenol composition) characteristics of honeys produced in BENIN's septentrion region. The physico-chemical parameter pH measured revealed the quality of honeys harvested in the septtemptrion. The pH ranged from 3.2 to 5.12 for the samples considered, and were therefore acidic. All the honey samples essentially comply with Codex Alimentarius quality standards for pH. Studies carried out by (Bogdanov et al 1999) reveal that nectar honeys have a pH between 3.5 and 4.5, while honeydew honeys have a pH between 5 and 5.5. Samples 1, 2, 3, 5, 6 and 8 have a pH in the range (3.5 and 4.5) and are therefore nectar honeys, while samples 4, 7 and 9 have a pH in the range (5 and 5.5) and are therefore honeydew honeys.

The total polyphenol composition of the honeys is interesting: our results clearly show that the sample (E5) is a very important source of phenolic compounds, and therefore has good antioxidant activity.

<u>**Outlook**</u>

Following these results, it would be advisable to carry out more detailed studies.

> ➢ Determine the free acidity of our samples, since acidity is an important quality criterion.
> ➢ Study the antioxidant activity of our honeys to assess the correlation between the quantity of total polyphenols and this activity.
> ➢ Research and assay of all types of polyphenols present in our honeys in BENIN
> ➢ Use cutting-edge techniques to identify the structures of our compounds, such as H.P.L.C. (High-Performance Liquid Chromatography) coupled with (MS) Mass Spectrometry and N.M.R. (Nuclear Magnetic Resonance), etc.

BIBLIOGRAPHICAL REFERENCES

1- **Alexandra Rossant** 2011 Honey, a complex compound with surprising properties p.133

2- **Aude Hyardin** 2008 Study of the food functionality of industrial dishes 238 pages

3- **Process Engineering" Department,** SPIN Center, Ecole des Mines de Saint-Etienne Page 19

4- **Beretta J., Giangiacomo G., Ferrero M., Orioli M. and Maffei Facino R., 2005.** Standarization of antioxidant properties of honey by a combination of spectrophotometric/fluorimetric assays and chemometrics. An. Chimica Acta ; 533 :185-191.

5- **Beta T., Nam S., Dexter J.E. and Sapirstein H.D., 2005.** Phenolic content and antioxidant activity of pearled ulreat and roller-milled fractions .Cereal Chem; 60: 390 -393.

6- **Bogdanov S., Lullman C. and Martin P., 1999.** International Honey Commission. Honey quality and international regulatory standards: review by the International Honey commission. Bee-World. 80(2) :61-69.

7- **Bruneau E.** Les produits de la ruche. In Le traité rustica de l'apiculture. Paris, Rustica, 2002, p. 354-384.

8- **Bruneton, J.**(2009). Pharmacognosie-Phytochimie,plantes médicinales, (4e éd),revue etaugmentée, Tec & Doc - Éditions médicales internationales,Paris, p 1288 .

9- **Catherine Mousinho, Marie-Christine Marinval.** Hives, apiaries and honey and wax harvests in France from the Middle Ages to the Modern era (13th-18th) 1-12

10- **C. Baillet** TP Spectrophotometry.doc

11- **Cimpoiu C., Hosu A., Miclaus V. and Puscas A.** 2012. Determination of the floral origin of some Romanian honeys on the basis of physical and biochemical properties. Spectrochimica Acta. Part A, Molecular and Biomolecular Spectroscopy ; 10 :1010-1016.

12- **Clémence HOYET** 2005Honey: from source to therapy p.106

13- **Dacosta, E.** (2003).Les phytonutriments bioactifs.Yves Dacosta (ed). Paris, **p317.**

14- **Delphine Ireland** 2010 Honey and its therapeutic properties Use in skin wounds 25 pages

15- **Dimitrova B., Gevrenova R. and Anklam E., 2007.** Analysis of phenolic acids in honeys of different floral origin by solid phase extraction and high performance liquid chromatography. hytochemicalAnalysis ; 18 : 24-32.

16- **Doukani Koula1*, Tabak Souhila1, Derriche Asma1, Hacini Zahira1** 2014 Physicochemical and phytochemical study of some types of Algerian honeys p.13 pages (**10**)

17- **Epifano F., Genovese S., Menghini L. and Curini M. 2007.** Chemistry and pharmacology of oxyprenylated secondary plant metabolites. Phytochemisty; 68:939 - 953.

18- **Ez-zohra Nkhili, 2009.** Polyphenols in Food: Extraction, Interactions with Iron and Copper ions, Oxidation and Antioxidant Power. PhD thesis in Food Science. University of Marrakech. Morocco. p.301

19- **Fatima Kholkhal** et al. 2013 Phytochemical study and evaluation of the antioxidant activity of Thymus CIliatus ssp. Coloratus 151 - 158

20- **Favier A. (2003).** Oxidative stress. Conceptual and experimental interest in understanding disease mechanisms and therapeutic potential. L'actualité chimique pp108-115.

21- **Gbèkponhami Monique Tossou1 Hounnankpon Yedomonhan1 Paulin Azokpota2 Akpovi Akoegninou1 Pablo Doubogan1 Koffi Akpagana3** 2011 Pollen analysis and phytogeographical characterization of honeys sold in Cotonou (Benin) 500-508

22- **Genot C., Eymard S., Viau M.** 2004,How to protect long-chain omega-3 polyunsaturated fatty acids from oxidation? Oléagineux, corps gras, lipides, vol. 11, n°2, p. 133-141.

23- **Georgé, S., Brat, P., Alter, P., and Amiot, M.J. 2005.** Rapid determination of polyphenols and vitamin C in plant derived product, J.Agric.Food Chem, 53: 1370-1373.

24- **Guerzou M. N., Nadji N.,** 2002. Comparative study between some local and imported honeys. Diploma in Agronomy. Université Ziane Achour de Djelfa- Algeria. Dissertation online

25- **Gutteridge, J.M. (1993).** Free radicals in disease processes: a compilation of cause and consequence, Free Radic Res Commun, 19:141-158.

26- **Halliwell, B; Gutteridge, J.M.C. (1999).** Free radicals in biology and medicine, Oxford, UK. **Hartmann T., 2007.** From waste products to ecochemicals: Fifty years research of plant secondary metabolism. Phyto chemisty ; 68: 2831 -2846.

27- **Hartmann T., 2007.** From waste products to ecochemicals: Fifty years research of plant secondary metabolism. Phyto chemisty ; 68: 2831 -2846.

28- **Hopkins, W.G. 2003.** Plant physiology. Edition Debock et lancier. P. 276.

29- **Hubert A.J. 2006.** Biochemical characterization and biological properties of soybean germ micronutrients. Study of its valorisation in nutrition and human health. Doctoral thesis in food quality and safety, Institut National Polytechnique de Toulouse. France p.174

30- **J.A. Djossou1, F.P. Tchobo1, H. Yédomonhan2, A.G. Alitonou1 & M.M. Soumanou1*** 2013Evaluation of the physicochemical characteristics of honeys marketed in Cotonou 163-169

31- **Kanoun Khadidja** 2011Contribution to the phytochemical study and antioxidant activity of *Myrtus communis* L. (Rayhane) extracts from the Tlemcen region (Honaine) p.118

32- **Kebieche Mohamed** 2009 Biochemical activity of flavonoid extracts from the *Ranunculus repens L* plant: effect on experimental diabetes and Epirubicin-induced hepatotoxicity p.143

33- **Khadhri Ayda1, El Mokni Ridha2 and Smiti Samira1** 2012-2013 phenolic compounds and antioxidant activities of two thistle extracts A GLU: Atractylis gummifera 44-52

34- **Knaggs, A.R.** (2003).The biosynthesis of shikimate metabolites.Natural Product Reports, 20: 119-36.

35- **Ladoh et al**. 2014 Antioxidant activity of methanolic extracts of Phragmanthera capitata (Loranthaceae) harvested from Citrus sinensis 7636- 7643

36- **Lavoisier,** 1987 Apiculture 6th edition Paris 570 p.

37- **Lefief-Delcourt A.** Le miel malin. Paris, Leduc.s, 2010, p.176

38- **Lesgards, J.F. (2000)**. Contribution à l'étude du statut antioxydant de l'homme; aspect chimique et biochimique. PhD thesis, 19-20.

39- **Maarouf A .2000.** Botanical dictionary Pp 129.

40- **Manallah Ahlem** 2012 Antioxidant and anticoagulant activities of olive pulp polyphenols Antioxidant and anticoagulant activities of olive pulp polyphenols p.132

41- **Marchenay P. and Berard L.** L'homme, l'abeille et le miel. Paris, De Borée, 2007, p.223

42- **Martin, S., Andriantsitohaina, R.**(2002).Mechanisms of cardiac and vascular protection of polyphenols at the level of the endothelium. Annales de cardiologie et d'angéiologie,

43- **Martos L., Ferreres F., Yao L., D'Acry B., Caffin N. & Thomás-Barberán F.A.** 2000, Flavonoids is monospecific Eucalyptus honey from Australia. J. Agric. Food Chem. **48**, 4744-4748.

44- **Meda A., Lamien C. E., Marco R. [et al.].** Determination of the total phenolic, flavonoid and proline contents in Burkina Fasan honey, as well as their radical scavenging activity. Food Chemistry, 2005, vol. 91, n°3, p. 571-577.

45- **Melle Kanoun khadidja** 2010/2011 Contribution to the phytochemical study and antioxidant activity of Myrtus communis L. (Rayhane) extracts from the Tlemcen region (Honaine) p.118

46- **Mlle Bellebcir Leila** 2007/2008 Study of phenolic compounds as biodiversity markers in cereals 119 pages

47- **N. Bentabet**, Z. Boucherit-Otmani, K. Boucherit 2014 Chemical composition and antioxidant activity of organic extracts of Fredolia aretioides roots from the Bechar region, Algeria 1-8

48- **Naczk, M., Shahidi, F.**(2004).Extraction and analysis of phenolics in food. Journal of Chromatography

49- **Oudjet kahina** 2012 HONEY A Commodity to Promote Infos-CACQE N°:003

50- **Oyaizu M., 1986.** Studies on products of browning reaction: antioxidative activity of products of browning reaction. J Nutr. Japon ; 44 :307-315.

51- **Paraskevi, Moutsatsou**(2007) The spectrum of phytoestrogens in nature: our knowledge is expanding. Hormones, **6** (3): 173-193.

52- **Pharmacie 1ere année** 2002-2003 Travaux pratiques de chimie analytique 13 pages

53- **Pham-Delegue M.-H.** Les abeilles. Genève, Minerva, p.1999, 206

54- **Philippe Morin** A new analysis technique: Spectrophotometry

55- **Pierron** 2008 pH measurement of aqueous solutions, dilution (EXAO) P. 10

56- **Pincemail ,J ; Defraigne, J.D. (2004).** Antioxidants: a vast network of defenses against the toxic effects of oxygen. Service de Chirurgie Cardio-vasculaire, Pro biox SA. Sart Tilman 4000 Liège, Belgium.

57- **Rabeharifara Zoelinoro Patricia** 2011 FOOD CHARACTERIZATION OF MALGARIC HONEYS WITH A VIEW TO AUTHENTIFICATION: CASE OF EUCALYPTUS HONEYS p.102

58- **Singleton V-L. and Rossi J-A. 1965.** Colorometry of total phenolics with pohsphomolybdicphosphotungstic acid reagents .American Journal of Enology and Viticulture ; 16 :44-158.

59- **Soman N. R., Baldwin S. L., Hu G. [et al.].** Molecular targeted nanocarriers deliver the cytolytic peptide melittin specifically to tumor cells in mice, reducing tumor growth. The Journal of Clinical Investigation, 2009, vol. 119, n°9, p. 2830- 2842.

60- **Tahraoui Fatima Zohra** 2013-2014 Contribution to the phytochemical study and antioxidant activities of Pituranthos scoparius (Guezzah) extracts by the iron reduction method: FRAP p.74

61- **Terrab A., Diez M.J . and Heredia F.J., 2003.** Palynological, Physicochemical and color characterization of Morocanhoneys : Orange (Citrus sp.) honey .International Journal of Food Science and Technology; 38 :387-394.

62- **Valko M., Rhodes C.J., Moncol J., Izakovic M., Mazur M. (2006).** Free radicals, metals and antioxidants in oxidative stress-induced cancer. Chemico-biological interactions. 160: 1-40.

63- **Voisenet (J.). -** Bestiaire chrétien : l'imagerie animale des auteurs du Haut-Moyen-Âge (Ve-XIe siècle). Toulouse: Presses universitaires du Mirail, 1994. 386 p. (Tempus collection).

64- **1st Year of Pharmacy** 2002-2003 Practical work in analytical chemistry Spectrophotometry Colorimetry

1. **Polyphenol biosynthesis**

Polyphenols are synthesized by two biosynthetic pathways (MANALLAH Ahlem 2012)

- That of shikimic acid, which after transamination and deamination leads to the cinnamic acids and their many derivatives, such as benzoic acids or simple phenols (Knaggs, 2003).

- Acetate, which leads to poly ß-coesters (polyacetates) of length leading by cyclization to polycyclic compounds such as dihydroxy-1,8 anthraquinones or naphthoquinones (Bruneton,1999 ;Naczk, et Shahidi,2004).

What's more, the structural diversity of polyphenolic compounds due to this dual biosynthetic origin is further enhanced by the possibility of simultaneous participation of both pathways in the elaboration of compounds of mixed origin, the flavonoids (Martin and Andriantsitohaina, 2002).

2. **Main groups of plant phytochemicals**

Polyphenols **are** considered almost universal compounds of plants. Structurally, they fall into several classes, ranging from compounds with a simple phenolic core (e.g. gallic acid) to complex polymeric compounds such as tannins. Polyphenols are the active ingredients of many medicinal plants. Generally speaking, they are found in all vascular plants, where they can be found in various organs: roots, stems, wood, leaves, flowers and fruits (Ez-zohra NKHILI 2009).

Table 4: Non-exhaustive classification of polyphenols :

Families	Classes	Subclass	Linear chemical formula
Polyphenols	Flavonoids	Flavones	C6-C3-C6
		Flavonols	C6-C3-C6
		Flavanones	C6-C3-C6

		Flavanols	C6-C3-C6
		Antocyanins	
	No flavonoids	Phenolic acid	C6-C1
		Stilbene	C6-C2-C6
		Lignane	(C6-C3)2
		Lignins	(C6-C3)n
		Water-soluble tannins	(C6-C3-C6)n
		Condensed tannins	(C6-C3-C6)2

3. <u>**Importance of polyphenols**</u>

- Role of natural antioxidants; Main mechanisms of antioxidant activity:

-Direct trapping of ROS (reactive oxygen species)

-Inhibition of enzymes involved in oxidative stress and chelation of trace metals responsible for ROS production

-Protection of antioxidant defense systems.

- Nutritional importance of polyphenols (phytomicronutrients); phenolic compounds play an important role in the organoleptic quality of fruit and vegetables, whether used fresh or after industrial processing.

- Veno-active" properties (flavonoids in particular): the ability to reduce the permeability of blood capillaries and strengthen their resistance (vitamin P action).

- Anti-allergic, anti-inflammatory, anti-ulcer, anti-cancer, etc. They are also used as additives in the food, pharmaceutical and cosmetics industries.

4. <u>**Polyphenol properties**</u>

- Hydrogen bond donors and acceptors (OH from phenol: acid)

- Anti-oxidant and free radical scavenging properties

- Chelation of metal ions (iron, copper, etc.)

- Solubility: polar solvents (depending on structure), e.g. glycosylated flavonoids are water-soluble. Organic solvents: methanol, ethanol, acetonitrile, isopropanol, nbutanol, ethyl acetate and diethyl ether (aglycone forms of flavonoids) acetone (anthocyanins).

5. <u>Main free radicals and comparison of their oxidizing power</u>

<u>Table 5</u>: Main free radicals and comparison of their oxidizing power(Kebieche Mohamed 2009).

Free radicals (nomenclature)	Chemical structure
Hydroxyl radical	$^{\cdot}OH$
Alkoxyl radical	$RO^{\cdot}$
Hydroperoxide radical	$HOO^{\cdot}$
Radical peroxide	$ROO^{\cdot}$
Nitric oxide radical	$NO^{\cdot}$
Alkoxyl radical	$RO^{\cdot}$
Hydrogen peroxide	H_2O_2
Peroxynitrite	$ONOO^{\cdot}$
Superoxide anion	$O_2^{\cdot-}$

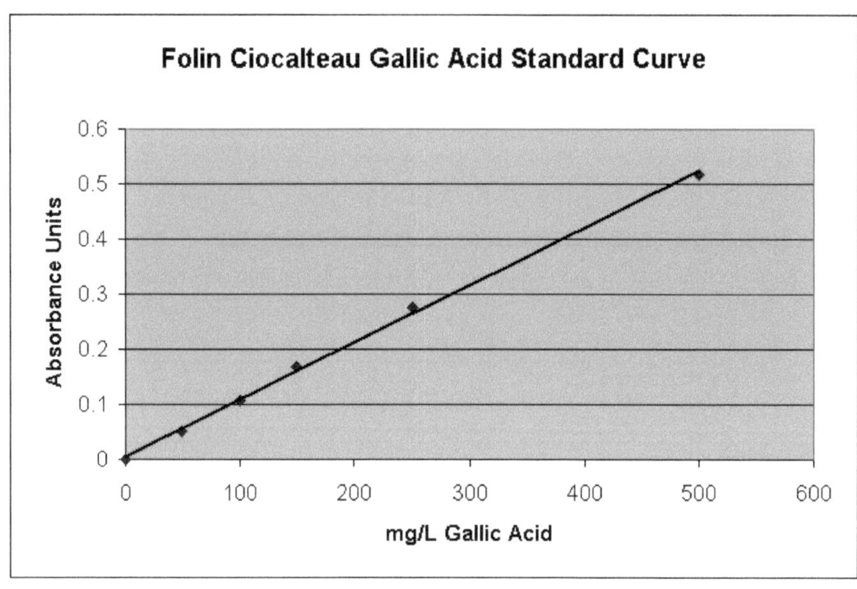

Figure 18 Gallic acid standard curve

I want morebooks!

Buy your books fast and straightforward online - at one of world's fastest growing online book stores! Environmentally sound due to Print-on-Demand technologies.

Buy your books online at
www.morebooks.shop

Kaufen Sie Ihre Bücher schnell und unkompliziert online – auf einer der am schnellsten wachsenden Buchhandelsplattformen weltweit! Dank Print-On-Demand umwelt- und ressourcenschonend produziert.

Bücher schneller online kaufen
www.morebooks.shop

Printed by Books on Demand GmbH, Norderstedt / Germany